SCIENCE NOTEBOOK

This notebook belongs to:

____